What Can I See... When I Look At Leaves?

Carole Diehl

To order additional copies of this book, contact:
Xlibris Corporation
1-888-795-4274
www.Xlibris.com
Orders@Xlibris.com

What Can I See...

...When I Look at Leaves?

Carole Diehl

For my daughter, Katie, (the Peeper)
with love,
Momma

Thanks to my husband, Jeff,
for his patience and understanding,

and special thanks to my friends, Mary Fuller and Frank Ammer,
for their suggestions and support.

What can I see when I look at leaves?

I can see different shapes.

Leaves are usually the flat, green parts of plants. They are typically green because they contain the pigment chlorophyll.
They come in a variety of shapes.

Some leaves are large and some are small.

Some leaves have smooth edges and some have jagged edges.

Some leaves are rounded and
some are pointed.

Some leaves are called simple leaves because they are made of only one part.
Compound leaves are made of many parts.

Some leaves are sharp like needles.

The shape of the leaf is one way to identify the plant.

I can see veins.

Leaves have veins that look like ridges or lines through them.
These veins carry liquid just like your veins do. Different kinds
of plants have different arrangements of veins.
Some veins look like branches and some veins look straight.
Veins carry water and minerals to all parts of the plant.

I can see different textures.

Leaves come in an array of textures.
Some are hairy and some are spiny.

Hairs and spines are modified parts of leaves. They help keep the plant from drying out and they also help protect the plant from being eaten.

I can see flowers.

Flowers are modified leaves, too. These leaves are called petals. They are colorful and smell good so that insects will land on them.

Insects drink the sugary liquid called nectar. Pollen, the yellow dust on the flowers, sticks to the legs of the insects and is carried to other plants. The pollen helps plants make seeds.

I can see yellow, red and orange colors.

Leaves have other pigments besides chlorophyll. In the fall, as the nights get longer, chlorophyll production stops. As the chlorophyll breaks down, it is not replaced. The other pigments that had been hidden by the chlorophyll can now be seen.

Some of the other pigments are red, yellow and orange.

I can see animals eating.

Leaves can make their own food by a process called photosynthesis. Their pigments have a special way of using the energy from the sun to make food for themselves and for other organisms.

Lots of animals, like sheep and buffalo, eat leaves.

Animals that eat only plants are called herbivores.
The antelope, giraffe and elk are herbivores.

They eat plants because that is the only way that they can get their own food.

I can see animals hiding.

The pigments in leaves not only help the plant make food, their colors can camouflage animals.

The inchworm and the katydid are two insects that hide among green leaves. The color of the leaves keeps them from being seen by animals that would like to eat them.

Other colors can also camouflage animals. This coyote is hidden by the grasses as it hunts for mice.

Animals, like butterfly larva and slugs, are easier to spot.

I can see places to live.

Many creatures make their homes in the leaves. The leaves give protection from the wind and the rain. They also hide the organisms from predators.

Tree frogs, birds and lizards often make their homes among the leaves.

I can see art.

Leaves can be used to make wonderful crayon rubs. You can make one, too.

Place a variety of leaves, vein side up, on a flat surface. Cover the leaves with a thin sheet of white paper. Rub the sides of crayons over the paper. The shapes and veins of the leaves should become visible.

A leaf collage can make a beautiful wall hanging.

Collect a variety of leaves. Press the leaves in a leaf press or place them on newspaper between heavy books or other weighty objects. Allow the leaves to dry for several weeks. Glue the dried leaves onto the paper of your choice, frame and hang.

What can I see when I look at leaves?

I can see shapes, textures and colors. I can see animals. I can see works of art.

What can you see...

when you look at leaves?

Vocabulary

Camouflage-natural way that animals blend into their surroundings in order to avoid being seen by predators

Chlorophyll-green pigment in plants that captures the light energy from the sun and uses it in photosynthesis

Compound leaf- leaf made of more than one blade, called leaflets, attached to a common stalk

Herbivore-animal that eats only plants

Needle-long, slender leaf of a pine tree

Photosynthesis- process where green plants turn carbon dioxide and water into carbohydrates and oxygen, using light energy trapped by chlorophyll

Pigments-molecules that absorb wavelengths in the visible range of light (red, orange, yellow, green, blue, indigo and violet)

Predator-animal that hunts, kills, and eats other organisms

Simple leaf-leaf having one blade attached to a stalk

Veins-conducting tubes that carry water and minerals throughout the plant

www.ingramcontent.com/pod-product-compliance
Lightning Source LLC
Chambersburg PA
CBHW050415180526

45159CB00005B/2282